Guide to

Eastern
Mushrooms

Written and Photographed
By E. Barrie Kavasch

hancock

house

D1546006

ISBN 0-88839-091-2
Copyright © 1982 E. Barrie Kavasch

Canadian Cataloging in Publication Data
    Kavasch, Barrie E.
        Field guide to Eastern mushrooms

        (Northeast color series)
        ISBN 0-88839-091-2

        1. Mushrooms - Canada, Eastern - Identifi-
        cation.* 2. Mushrooms - Northeastern
        States - Identification. I. Title. II. Series.
        QK617.K37   589.2'223'0971
        C81-091042-X

Editor Margaret Campbell
Production Peter Burakoff
Layout/Design Diana Lytwyn and Linda Rourke
Typeset by Sandra Sawchuk and Anne Whatcott in Megaron type on
an AM Varityper Comp/Edit

Printed in Canada by Friesen Printers

Published by

Hancock House Publishers
256 Route 81, Killingworth, CT, U.S.A. 06417
Hancock House Publishers Ltd.
19313 Zero Avenue, Surrey, B.C., Canada V3S 5J9

Table of Contents

Acknowledgments

Culinary experimentation and partial research on many of these species was carried out at the American Indian Archaeological Institute in Washington, Conn. My personal appreciation goes to COMA, the Connecticut Mycological Association, and CVMS, the Connecticut Valley Mycological Society, and their many helpful and interested members. Special thanks to Mary Anna Wray, Raelene Gold, Dodie Nalven, Vincent Marteka, Dr. Barry Wulff, Sylvia Steine, Gary Lincoff, Elizabeth Jensen, and to David Hancock for bringing this to fruition. Special appreciation to Dr. Barry Wulff for reading this manuscript.

A number of these fungi were photographed *in situ* in the Native Plant Arboretum of the American Indian Archeological Institute, and on the grounds of the White Memorial Foundation in Litchfield, Connecticut.

My appreciation to John Pawloski for a number of his excellent mushroom photographs included herein, and to Jean Murkland for manuscript help and typing.

Author's Profile

E. Barrie Kavasch is a self-taught naturalist and ethnobotanist with years of experience and special interests in the plant world. She is an artist and photographer, and is also the author of *Native Harvests,* and *Botanical Tapestry.* She teaches and lectures throughout New England and demonstrates the harvesting and preparation of "wild edibles."

John Pawloski photo

A puff ball "puffing" its spores.

John Pawloski photo

4

Introduction

Throughout time, mushrooms and related fungi have fascinated people all over the world. American Indians used the mushrooms on this continent for thousands of years, and descriptive ethnomycology has come to us from American Indians, Asians, Chinese, Africans, and Europeans.

During various periods wild mushrooms have been a staple food in many cultures, and they offer important minerals and some vitamins not readily accessible in other foods. However, the hallucinogenic and deadly properties of some species are in sharp contrast to the delicious edibility of many other wild species.

The full importance of fungi in the plant world and to mankind is only partially understood. Some mushroom species are saprophytes, existing on dead organic matter and assisting the decomposition process. Others are a source of drugs, notably the diminutive fungus *Claviceps,* which is a parasite on rye grass, and from which numerous alkaloids and chemicals, including LSD, are derived. The wonder drug penicillin, too, was first extracted from the fungus *Penicillium,* a green mold. And for countless centuries the yeast fungi have been used to ferment beverages and enhance food. Mushrooms have some vital roles to play.

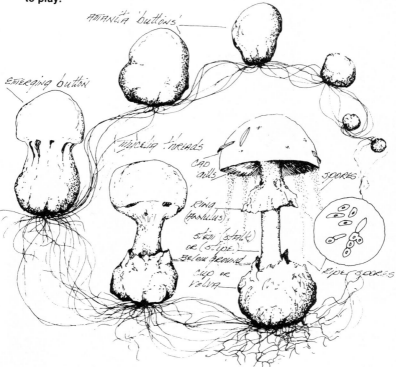

The life cycle of *Amanita virosa.* Destroying Angel (Woodland mushroom) development from underground buttons, mycelia from spores (circle), to the mature gilled mushroom.

What is a Mushroom?

Mycologists, scientists who study fungi, generally include all fleshy fungi under the category 'mushroom,' even the more diverse fungi like the puffballs, the spongy boletes and the brain-like morels. The old colloquial term "toadstool" has no special meaning in the serious study of mushrooms.

How can you tell if it is a toadstool or a mushroom?

John Pawloski photo

It is only a toadstool if you find a toad sitting on it.

Mushrooms belong to the world of non-flowering plants, within the large primitive group known as fungi. Botanists estimate that more than 300,000 species of fungi exist on earth. What we recognize as a mushroom is the fruit of a hidden plant that is essentially a network of root-like threads spreading within wood, dung, soil, or layers of leaves. This fine fish net of roots is called a *Mycelium*. The *Mycelia* obtain food by releasing chemical enzymes into the surrounding soil or wood. When enough food and moisture have been absorbed, and when circumstances are best for the production and dispersal of spores, the mushroom begins to emerge, or "flush." This is usually after a soaking rain.

Mushrooms do not contain chlorophyll. To obtain energy, many must form beneficial partnerships with herbs, shrubs and trees. Others are parasitic on plants and animals and even other mushrooms.

Identification

Of the more than 800 known species of mushrooms in New England (the northeast), a few are so familiar and distinctive that you can recognize them easily by sight, but others are so difficult that experts cannot agree on their identity. If you have problems identifying a mushroom, don't despair. Many species require chemical tests and microscopic analysis for final identification.

As a first step to knowing your mushrooms, however, you should make detailed notes, as well as spore prints, for every specimen collected. Examine each mushroom systematically, beginning with the cap and working toward the base. Note any unique odor or change in color when handled. Odor is often difficult to describe or highly variable, so it has been included in the descriptions in this book only when it is particularly distinctive. Cut one specimen in half, note the color of its flesh and check for insect predators. Examine the gill or pore surface, and note the color, thickness, and attachment of the cap to the stalk.

Because raw flavor may be quite different from cooked flavor, many mushroom collectors make tentative taste tests in the field. *They do not eat mushrooms raw, but taste for diagnosis and then spit them out.* It is not too different from the technique of wine tasting.

One of the most important steps for correct identification is to make a spore print. The tiny, single-cell spores serve the same purpose as the more complex seeds of higher plants. To prepare a spore print, remove the stalk very close to the gills or pores. Place the gill or pore surface on white or gray paper. Enclose it in a waxed paper bag if you are in the field, or cover it with a glass or bowl if you are home. It will take anywhere from half an hour to overnight for some specimens to drop their spores. The spores are finer than chalk dust and just as elusive. In order to preserve a nice spore print for reference or artwork, you might cover it carefully with contact paper. It is a bit like trying to preserve a spider's web.

  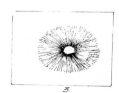

Making the spore print: 1) Cut stem off close to cap.   2) Place cap gills down on card or piece of paper. Cover with glass or paper for several hours to overnight.   3) Lift cap off gently and discard. Spore print is important for color and arrangement.

Collecting Mushrooms

Most mushrooms fruit or "flush" seasonally. Certain conditions, such as soaking rain, cool nights, and warm days, can produce abundant flushes of common mushrooms.

Mushroom collectors require a large, flat mushroom basket, a sharp knife, waxed paper bags (in order to keep different species separate from one another), cards or paper for spore prints, paper for notes and labels, a good guide book, and a hand lens.

Chipmunks and squirrels also harvest the seasonal flush of mushrooms. You will find large gilled mushrooms standing partially eaten in the woods, distinct teeth marks showing in the remains. Occasionally you'll find stems and cap pieces drying in tree boughs for later storage. *However, the fact animals eat a certain type of mushroom does **NOT** indicate edibility for humans.*

After careful identification, edible species should be checked for insects and small predators. Brush off dirt and cut off the stalk base before filling bags with choice specimens. This care in the field will assist later cleaning and preparation.

Mushroom societies and clubs have been formed in many areas by interested people who wish to increase and share their knowledge of mushrooms. They usually sponsor local mushroom hunts, or forays as they are sometimes called. If you are truly interested in knowing more about wild mushrooms, you should consider membership in one of the mushroom societies in your area, or NAMA, the North American Mycological Association, Portsmouth, Ohio, 45662.

Mushroom spores are tiny, single-cell creations, much smaller than the more complex seeds of higher plants. Easily airborne, spores can be distributed over great distances. This ensures that some of the spores will find areas that have the required characteristics for mushroom growth.

Mushrooms in the Kitchen

The cultivated mushroom *Agaricus bisporus* is edible and enjoyed raw, but wild mushrooms should not be eaten raw. Some are acutely indigestible and contain toxins which are destroyed by cooking. Also, individual body chemistry varies so considerably that some people might not be able to digest a well-prepared wild mushroom, while others would have no trouble at all. Always taste sparingly until you develop your "wild palate."

Most collected mushrooms should be refrigerated until they can be prepared for eating. The *Coprinus* (Inky Cap) group, however, must be cooked as soon as possible in order to arrest their tissue-dissolving enzyme. They should be carefully wiped, washed and dried, cut in half, and then simmered for four to six minutes in a small amount of butter. I prefer to use an iron skillet for all my mushroom preparations.

Usually the cap, upper stalk, and gills or pores provide the best eating. Cut each mushroom in half to check for internal predators. Keep your use of water for washing to a minimum, as the fleshy mushrooms will absorb it and become watery when cooked.

It is best to gently sauté each new specimen alone in a small amount of butter or margarine until it is just barely shiny or translucent. Then taste sparingly and savor its unique flavor in order to determine how you want to season and serve it. I also sauté mushrooms before I freeze them.

Mushrooms may also be prepared as hors d'oeuvres, relishes or pickles, and may be added to gravies, soups, salads, sauces, vegetables, pastas, casseroles, meat dishes and omelettes. Certain mushrooms become accents in breads, pastries, and cookies, and a few lend themselves to candying.

Mushrooms can be dried, canned, frozen or powdered, for future use. I prefer to freeze or dry my favorites. The corals and jellies as well as the gilled mushrooms dry very well if placed on clean foil sheets in a warm oven overnight. Each species should then be sealed separately in an air-tight glass jar or plastic bag along with the date and label. These dried specimens can be easily revived in winter soups or stews for delightful taste accents.

Warnings

1. Do not eat wild mushrooms raw. Some may be tasted when raw in order to check field characteristics, but then spit them out.
2. Use only specimens in good condition. Check for insect damage, mold, and old age (decay).
3. Cook or preserve wild mushrooms as soon as possible after harvesting.
4. When eating a new species for the first time, taste only a small amount at the first meal to be certain you can digest it well. Do not gorge on a new specimen.
5. Keep all species separated from one another while collecting. Use paper or waxed paper (not plastic) bags.
6. Do not mix different mushrooms species in one meal.
7. Do not eat any specimens you are not *absolutely* sure of.
8. Do not use pictures alone as your guide. Study all the field characteristics of each species, especially the spore print color.
9. Proceed cautiously until you ascertain your own tolerances. Body chemistries and metabolisms vary widely from person to person.

GILLED FUNGI

The *Amanitas* are beautiful to study, but when handling them you must be certain *not to mix them* with possible edible species. Just the spores from a poisonous *Amanita* which might cling to a cooked edible fungi would be enough to cause severe gastro-intestinal upset or death. *Amanitas* account for ninety percent of all deaths due to mushroom poisoning.

Amanitopsis is an old genus name for *Amanitas* that are without a ring. All *Amanitas* start as oval 'buttons' without a tissue veil.

Amanita gemmata

Cap color: Yellow with white warts
Edibility: Poisonous
Size: To 12 cm. broad
Spore print: White
Shape: Convex to flat and slightly depressed in age
Gills: Free, close, narrow, and white
Texture: Viscid and smooth
Flesh: Soft, thin, and white
Stalk: Tall and sturdy, smooth above ring, has a basal bulb
Habit/habitat: Single with infrequent groups under mixed hardwoods and conifers
Distribution: Widespread
Seasons: Spring, summer and fall
Odor: Mild

This tall, fragile *Amanita* hybridizes with *Amanita pantherina* (according to Miller, 1978): "producing a series of forms between the two." It is best not to handle or collect either of these.

Amanita pantherina   Panther agaric

Cap color: Brown with pointed white warts
Edibility: Deadly poisonous
Size: To 12 cm. broad
Spore print: White
Shape: Convex, flattening in age
Gills: White, crowded, free, and edges are scalloped
Texture: Viscid with soft warts
Flesh: Thin and pale
Stalk: Slender, tall and white, gradually enlarging to an egg-shaped basal bulb
Habit/habitat: Several to many on ground beneath mixed hardwood-conifer woods
Distribution: Widespread
Seasons: Spring, summer and fall
Odor: Not unpleasant

This is one of the few deadly poisonous mushrooms and should not be handled or collected.

Amanita virosa
Destroying angel

Cap color: Pure white
Edibility: Deadly poisonous
Size: To 9 cm. broad
Spore print: White
Shape: Egg-shaped, conic to convex, to flat with maturity
Gills: White, close, narrow and free
Texture: Viscid to sticky when wet, smooth
Flesh: Firm and white
Stalk: White, tall, sturdy (with showy white skirt), enlarging to bulbous volva
Habit/habitat: Usually solitary but frequently seen
Distribution: Widespread, common in eastern North America
Seasons: Spring, summer and fall
Odor: Odorless

This showy pure white mushroom is a common eyecatcher in northeastern woods. It is collected photographically but otherwise should not be touched.

Amanita vaginata
Grisette

Cap color: Pale yellow-brown to gray
Edibility: Non-poisonous / edible?
Size: To 9 cm. broad
Spore print: White
Shape: Conic to nearly flat with a slight knob
Gills: White, close without warts
Texture: Viscid, thin and soft
Flesh: White, thin and soft
Stalk: Tall and slender, white and smooth, no veil, enlarging to a bulb at base (white sac-like volva)
Habit/habitat: Single or scattered under mixed woods
Distribution: Widely noted
Season: Spring, summer and fall

Some mycophagists note this one as edible but I cannot recommend it.

Amanita flavoconia

Cap color: Orange with yellow warts
Edibility: Poisonous
Size: To 8 cm. broad
Spore print: White
Shape: Opening to broadly convex
Gills: White, close and free
Texture: Viscid
Flesh: White to cream, thin and firm
Stalk: Yellow to white, tall and sturdy enlarging to a basal bulb
Habit/habitat: Usually single to a few under mixed woods.
Distribution: Noted throughout New England and Canada
Season: Spring and fall

This is a striking fungi to find in the woods, and closely resembles *Amanita muscaria,* the poisonous "fly agaric." Neither one should be handled if you are seeking and picking edible mushrooms. *Do not collect.*

Amanita citrina

Cap color: Pale green
Edibility: Poisonous
Size: To 10 cm. broad
Spore print: White
Shape: Convex with a broad knob
Gills: White, close and free
Texture: Viscid and cottony to smooth
Flesh: Soft and white
Stalk: White, tall and sturdy, smooth to rough below skirt, enlarging to a bulb at the base
Habit/habitat: Single to several scattered in mixed woods
Distribution: Seemingly abundant in the east
Season: Late summer through late fall

This slender mushroom identifies itself with its characteristic *Amanita* features: convex, sometimes warted cap, distinctive skirt (ring), and enlarged basal bulb just beneath the ground. This one and *Amanita virosa* are usually plentiful right after a good drenching rain.

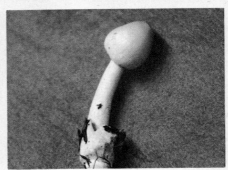

Lepiota procera
Parasol mushroom

Cap color: Tan with brown scales
Edibility: Edible, choice, mild taste
Size: To 24 cm. broad
Spore print: White
Shape: Convex to broadly convex, low knob
Gills: Close, broad, even, free, white
Texture: Dry, soft, fibrous quality
Flesh: Soft, white, non-bruising
Stalk: Very tall (to 40 cm.) and slender (sturdy), enlarging to small bulb and base with hairy ring (collar)
Habit/habitat: Several to scattered in weeds, brush, beneath mixed conifer and deciduous woods
Distribution: Widespread
Season: Spring and fall
Odor: Mild and pleasing

This is one of our choicest edible species, with impressive caps, which are the best part. These are delicious cut into strips (julienned) and sautéed in sunflower seed oil with your favorite herbs, to taste.

Lepiota cepaestipes

Cap color: Whitish to buff
Edibility: Edible, mild taste
Size: Up to 60 mm. broad
Spore print: White
Shape: Conic to nearly flat
Gills: White, free and crowded
Texture: Scaly and powdery
Flesh: White and soft
Stalk: White to buff, smooth to minutely hairy, enlarging to a small basal bulb
Habit/habitat: Several to clusters on wood chips or loam
Distribution: Eastern North America
Seasons: Spring, but usually summer and fall
Odor: Mildly pleasant

This species is tasty sautéed in butter, without its sturdy stem. Because it is so mild, it readily takes on the flavors of spicy sauces and herbs, so season to taste.

The *Hygrophorus* **is the most colorful group of gilled fungi, though a number of species can be drab. They all have thick waxy gills attached to the stalk, and they all grow on duff or on the ground. There are no known poisonous species in this group, and the** *Hygrophorus* **may be substituted for commercial mushrooms in almost any type of recipe.**

Hygrophorus coccineus

Cap color: Blood red
Edibility: Edible
Size: To 5 cm. broad
Spore print: White
Shape: Conic, convex with a low knob developing with age
Gills: Broadly attached, close, yellowish orange, waxy
Texture: Moist to tacky
Flesh: Soft, breaks easily, reddish orange
Stalk: Tall, thin, striate to twisted, moist, red to orange, no veil, and yellowish around base.
Habit/habitat: Scattered to a few on soil or grass beneath hardwoods and conifers
Distribution: Widely noted
Season: Late summer and fall

Hygrophorus russula

Cap color: Pink-streaked, with brownish-purple hairs
Edibility: Edible
Size: To 12 cm. broad
Spore print: White
Shape: Convex, smooth
Gills: Attached (adnate) to extending part way down stalk
Texture: Waxy and dry, but viscid in wet weather
Flesh: Thick, firm, white to pink
Stalk: Long, evenly thick, dry, white, no veil
Habit/habitat: Scattered to grouped under oaks
Distribution: Widespread but mostly in the east
Seasons: Late summer and fall

This species is noted to have a good flavor and is excellent sautéed with herbs, butter and eggs.

Hygrophorus flavescens

Cap color: Brilliant yellow-orange
Edibility: Non-poisonous
Size: To 7 cm. broad
Spore print: White
Shape: Broadly convex, ruffled edges
Gills: Waxy yellow, notched, and close together
Texture: Viscid to dry
Flesh: Yellowish and thin
Stalk: Pale yellow, tall and slender, no veil
Habit/habitat: Scattered and numerous on soil or moss in mixed woods
Distribution: Widely known
Seasons: Spring, summer and fall

Hygrophorus marginatus

Cap color: Brilliant lemon-yellow
Edibility: Non-poisonous
Size: To 5 cm. broad
Spore print: White
Shape: Convex to flat with a low knob
Gills: Yellow and waxy
Texture: Smooth and moist to dry
Flesh: Same color as cap and very thin
Stalk: Pale yellow to whitish, tall and slender, no veil
Habit/habitat: Single to several on soil under mixed woods
Distribution: Widely known
Seasons: Spring and fall

Both of these species are tiny, highly colorful, gilled fungi, usually found on moss in damp woods.

M. Morris photo

The genus *Lactarius* belongs to the "milk-mushroom" group. They contain a milky latex which is easily found by cutting across the gills, or breaking off a piece of the cap. There are different colors of latex in different species, and the color noted is important for correct identification.

Mycorrhizae, the inter-relationship between mushroom threads and a tree's roots, is a mutually beneficial relationship. Many mushrooms form *Mycorrhizae* only with certain trees. The *Lactarius* forms such unique partnerships with conifers and hardwoods.

Lactarius rimosellus

Cap color: Apricot-red and cracked (rimose) when aged
Edibility: Edible, mild flavor
Size: To 40 mm. broad
Spore print: White
Shape: Convex to slightly depressed-convex when aged
Gills: Close, extending down stalk a short way, and white to creamy yellow
Texture: Dry, smooth, without hairs
Flesh: Pale to same color as cap; latex is white, unchanging
Stalk: Long and sturdy, smooth, dry with a slightly powdery feeling
Habit/habitat: Scattered to grouped on ground near or under conifers
Distribution: Widely noted
Seasons: Summer and fall
Odor: Aromatic, like sweet clover

This is a particularly distinctive mushroom. These specimens are very aromatic, even when dried. Caution is urged with this group as some of the *Lactarius* have known toxins, and there are near look-alikes which range from non-edible to poisonous.

Lactarius subdulcis

Cap color: Amber-orange with red tints
Edibility: Edible, tastes acrid and increases in bitterness with time
Size: To 5 cm. broad
Spore print: White
Shape: Convex to flat, depressed in center, and has a small knob at maturity
Gills: Close, light to flesh-colored, and extend partially down stalk
Texture: Waxy, moist, smooth, hairless
Flesh: White, not bruising; latex white, unchanging
Stalk: Thick, sturdy, smooth, hollow with age, pale to same color as cap
Habit/habitat: Scattered to grouped on forest floors
Distribution: Widely noted, especially in wet areas in conifer woods
Seasons: Spring, summer and fall
Odor: Mild, not distinctive

This lovely *Lactarius* is known as the candy mushroom on the west coast and according to Miller (1978), it ". . .is used to make persimmon and candy mushroom pudding."

Peter Katsaros photo

Russula emetica   Emetic russula

Cap color: Bright red, fading in age
Edibility: Non-edible, possibly poisonous, instant strong acrid taste
Size: To 12 cm. broad
Spore print: White
Shape: Broadly convex
Gills: Adnate, close, sometimes forked, white
Texture: Viscid, smooth and hairless, margin striate
Flesh: Very soft, white, thin; pink just beneath skin
Stalk: Long and thick, dull white, hollow at maturity, no ring
Habit/habitat: Scattered to grouped on ground or in moss
Distribution: Widely found
Seasons: Summer and fall
Odor: Mild

This lovely *Russula* **has been cooked and eaten, but it is not recommended. There are many red** *Russula* **in North America. Because they are difficult to know, they are best collected photographically.**

Russula paludosa

Cap color: Orange-red, smooth, hairless
Edibility: Edible, tastes mildly acrid
Size: To 14 cm. broad
Spore print: Yellow
Shape: Convex to nearly flat or depressed when aged
Gills: Notched (adnexed), separated, warm buff color
Texture: Viscid and smooth
Flesh: Soft and white
Stalk: Long and evenly thick, dry, dull white, smooth
Habit/habitat: Scattered to grouped under mixed hardwoods and conifers
Distribution: Common throughout North America
Seasons: Summer and fall
Odor: Mildy fishy

This abundantly flushing *Russula* **is edible, but not choice. A certain amount of confusion can be caused by numerous near look-alikes with subtle differences.**

B.L. Wulff photo

Peter Katsaros photo

In the complex web of life, many of the fungi are re-cyclers of dead organisms; others prey on living hosts, thereby helping to regulate animal and plant populations. There are thousands of fungi that affect our lives: yeasts, breadmolds, wheat rust, mildew, athlete's foot fungus, and ringworm are only a smattering of the more common ones.

Russula xerampelina   Woodland russula

Cap color: Purplish red
Edibility: Unknown
Size: To 16 cm. broad
Spore print: Yellow
Shape: Broadly convex to depressed

Texture: Viscid when young, dry when aged
Flesh: Firm and white
Stalk: Long and evenly thick to flaring at base, hollow with age, white to flushed pink
Habit/habitat: Scattered or grouped under conifers
Distribution: Widely noted
Season: Summer
Odor: Distinctive lobster smell, strong when dried

This is reported as edible in Europe, but is not favored in the United States.

Lentinellus ursinus   Bear mushroo

Cap color: Whitish to creamy
Edibility: Non-poisonous, tastes extremel bitter
Size: To 10 cm. broad
Spore print: White
Shape: Convex, curled edges to nearly fla shelf-like layers
Gills: Close, pinkish, ragged saw-toothe edges
Texture: Top of cap at attachment is hairy
Flesh: Firm, white to buff
Stalk: No stalk
Habit/habitat: Several overlapping clusters o stumps and logs of hardwoods and conifers
Distribution: Widely known
Seasons: Summer, fall and winter, during coo moist periods
Odor: Fruity

This is a distinctive, showy, non-edible species.

Peter Katsaros photo

Pleurocybella porrigens Angel wings

Cap color: White to creamy
Edibility: Edible, tastes mild and agreeable
Size: To 30 cm. broad
Spore print: White
Shape: Broad, convex, petal-like fans
Gills: White to creamy, crowded, narrow, and radiating out from beneath base
Texture: Dry and minutely hairy
Flesh: White, pliable and very thin
Stalk: No stalk
Habit/habitat: Layered in overlapping clusters on conifer stumps and logs, especially hemlock
Distribution: Widely noted
Seasons: Late summer and fall
Odor: Mildly pleasant

The younger bodies are best. Check among gills carefully to be sure there are no insects or larvae housed within those you harvest for eating.

Pleurotus sapidus Purple-spored fan

Cap color: Dull white to tawny brown
Edibility: Choice, edible, tastes deliciously mild to slightly spicy
Size: To 30 cm. broad
Spore print: Lilac
Shape: Convex, shell-shaped fans; margins sometimes wavy
Gills: White, broad, and well separated, radiating out from blunt base of attachment
Texture: Smooth, moist, hairless
Flesh: White, thin, and pliant
Stalk: Short to absent, off-center
Habit/habitat: Single to numerous in overlapping clusters in mixed woods
Distribution: Wide
Seasons: Spring, early summer, fall and winter
Odor: Fragrant, fruity, almost of anise

This delicate mushroom flushes during cooler weather. It is closely related to the notable "oyster mushroom" *(P. ostreatus)* **which is very similar in growth and shape, having a white to buff spore print. These delicious mushrooms play hosts to shiny black beetles which live between their gills, therefore careful checks after harvesting and before preparation are essential. It is not unusual to find either of these fan-shaped mushrooms in the northeast during a winter thaw. Relish these ghostly edibles as "fresh-frozen" winter delight!**

Armillariella mellea Honey mushroom

Cap color: Honey colored, quite variable
Edibility: Choice, edible, tastes mild
Size: To 12 cm. broad
Spore print: White
Shape: Convex-knobbed to flat and depressed
Gills: White to cream, rusty stains with age
Texture: Dry and hairy, sticky when wet
Flesh: White and fairly thin
Stalk: White to buff, sturdy to thickly robust, cottony veil, sometimes absent
Habit/habitat: Single (rarely) to large cespitose (tight finger-like) clusters on wood, or on soil over buried wood
Distribution: Widely known, often recurring in same spot
Seasons: Summer and fall
Odor: Fresh and mildly woodsy

These distinctive, robust mushrooms are extremely variable in cap color, and one of our prized edibles. It is possible to gather "honeys" by the bushel in late September. They are excellent sautéed or pickled and canned. They preserve and freeze well after preparation.

Armillariella mellea **are associated with "shoestring root rot," a virulent parasite. This fungus causes extensive damage to hardwoods, especially oaks. But its presence in our woods is an old one and further study is necessary to prove its detrimental aspects within the life-cycle of the forest.**

Catathelasma ventricosa

Cap color: Ash gray to whitish
Edibility: Good, edible, tastes mildly spicy
Size: To 11 cm. broad
Spore print: White
Shape: Convex to broadly convex, and umbrella-like
Gills: Pinkish-buff, close, and extended partially down stalk
Texture: Smooth hairless, and dry
Flesh: White and firm
Stalk: White to dull yellow, tall, sturdy, tapering to a narrow base, double veil
Habit/habitat: Several to cespitose on wood
Distribution: Eastern North America
Seasons: Late summer and fall
Odor: Mild to odorless

These fleshy caps and diced stems are delicious when sautéed in butter. These fungi must be checked thoroughly for they have many tiny predators that quickly invade their stalks and caps.

rmillaria zelleri

ap color: Orange-brown
libility: Edible
ze: To 15 cm. broad
ore print: White
ape: Convex with a broad knob
lls: White to buff, crowded and attached
xture: Viscid
esh: White, thick and bruises orange-brown
alk: Whitish, thick and tapering to a blunt
se
abit/habitat: Scattered and cespitose under
rdwoods
stribution: Widely spread
asons: Autumn

**is sturdy fungus is excellent when diced
d sautéed in butter and seasoned with
orcestershire sauce. They are delicious
ough to take the place of meat as a main
urse.**

Armillaria caligata

Cap color: Reddish brown to burnt umber
Edibility: Choice, edible
Size: To 12 cm. broad
Spore print: White
Shape: Broadly convex
Gills: White, close and broad
Texture: Dry and smooth with a cottony ring
Flesh: White and firm
Stalk: Long and thick, tapering to a pointed
base, creamy white
Habit/habitat: Scattered to clustered under
mixed woods
Distribution: Principally eastern North
America
Season: Autumn

This *Armillaria* **is very similar to the previous
one in size and preparation. A few good
specimens can fill a family's dinner needs,
providing a steaming, delicious vegetable
dish. Check them thoroughly for insect and
snail infestation, and wipe caps and stems
clean. Do not soak mushrooms in water, as
they absorb like a sponge. It's better that they
absorb the butter and sauce.**

Marasmius oreades Fairy ring mushroom

Cap color: Light tan to reddish brown
Edibility: Choice, edible, tastes delicious and mild
Size: To 6 cm. broad
Spore print: White to buff
Shape: Convex to bell-shaped
Gills: Light buff, adnexed, fairly well separated and broad
Texture: Dry, smooth and hairless
Flesh: White, thin and watery
Stalk: Buff to brown, tall and slender, tough and dry, no veil
Habit/habitat: Grows in groups or fairy rings in grassy habitats
Distribution: Widely found, common
Seasons: Spring, summer and fall
Odor: Fragrant

Fairy ring colonies increase in area annually as the advancing *mycelium* **(fungal threads) expand their perimeters and interior areas receive less nourishment. This is a unique characteristic of these choice little mushrooms. Their cooked flavor enhances that of other vegetables. However, great care should go into identification as there are similar small brown mushrooms occurring in lawns that are poisonous.**

Marasmius scorodonius

Cap color: Brown with red tints and whitis margins
Edibility: Non-poisonous
Size: Tiny to 12 cm. broad
Spore print: White
Shape: Convex to flat
Gills: Whitish, close, narrow and attached
Texture: Dry
Flesh: Whitish, and very thin
Stalk: Tall and thin, tough, smooth, hairless white tapering to black
Habit/habitat: Groups growing on twigs and i grass
Distribution: Widely noted
Seasons: Spring, summer and fall
Odor: Strong odor of garlic when cap is crushe

There are no reported toxins or poisons i this family, but their thin toughness make them undesirable as edibles.

Clitocybe gigantea

Cap color: White to buff
Edibility: Eaten infrequently, mild taste
Size: To 45 cm. broad
Spore print: White
Shape: Very broad, convex to flat to deeply depressed and funnel-shaped
Gills: White, crowded, and extending down stalk
Texture: Cottony, moist to dry
Flesh: White and firm
Stalk: Same color as cap, very thick and sturdy, smooth and dry, no veil
Habit/habitat: Scattered or in fairy rings in open mixed woods
Distribution: Widely found but not common
Seasons: Late summer and fall
Odor: Faint woodsy fragrance

This is a large, showy, forest mushroom of late summer which, when found fresh, can be collected in great amounts. Though sizeable, it is not, to my knowledge, a choice edible. It does dry well for preservation for herbariums and scientific study.

Clitocybe odora   Anise clitocybe

Cap color: Bluish green to greenish gray (quite variable cap color)
Edibility: Edible
Size: To 10 cm. broad
Spore print: Pinkish to cream
Shape: Convex to broadly convex, slightly depressed in center
Gills: White to pale buff, adnate to extending down the stalk
Texture: Moist and streaked radially with hairs
Flesh: White, soft and thin
Stalk: White to buff, moist, thick and sturdy
Habit/habitat: Scattered beneath mixed hardwoods and conifers
Distribution: Widely found, common
Seasons: Summer and fall
Odor: Distinctively anise in young specimens

The cap color is quite variable in this common woods mushroom, but its fragrance of anise is consistent. This fungus lends charm and flavor to oriental and middle-eastern dishes, and complements rice and vegetables beautifully.

Peter Katsaros photo

Clitobybe nebularis

Cap color: Gray to toast brown
Edibility: Edible, uncooked it has a disagreeable taste
Size: To 15 cm. broad
Spore print: Pale yellow
Shape: Broadly convex, flat, edges fluting upwards with incurved margins
Gills: Whitish, close, attached to stalk
Texture: Dry and smooth with radiating flat-lying hairs
Flesh: White, thick and tough, moist
Stalk: Tall and thick with enlarged base, silky white, no veil
Habit/habitat: Solitary to several on soil under conifers
Distribution: Known on both coasts and in Canada, common
Seasons: Summer and fall
Odor: Disagreeable

This large, robust, succulent-looking mushroom is enough, alone, to feed one person. Pan-fried and mixed with onions and your favorite herbs, its disagreeable raw taste dissipates and it is delicious. *C. robusta* **is very similar to** *C. nebularis* **except its cap and stalk are white to buff.**

Clitocybe nuda

Cap color: Lavender-gray
Edibility: Choice, edible
Size: To 15 cm. broad
Spore print: Pinkish buff
Shape: Broadly convex, flat with uplifted margin and sometimes a low knob
Gills: Pale lilac, adnexed, brownish when aged
Texture: Smooth, dry and hairless
Flesh: Light lilac to buff and thin
Stalk: Pale lilac, short, slender enlarging to an oval bulb at the base
Habit/habitat: Single to several on woodland floor and grassy areas
Distribution: Common and widely noted
Seasons: Summer and fall
Odor: Fragrant

This is a delicious fungus with quite a range of color variations in the cap. This one can be found in large quantities, and is excellent when sautéed in a small amount of butter. Its lovely fragrance enhances its fine, mild taste.

ymnopilus spectabilis
Dyer's fungus

ap color: Yellow-orange
dibility: Non-poisonous, not eaten
ize: To 18 cm. broad
pore print: Rusty orange to orange
hape: Convex to nearly flat
ills: Attached to and extending down stalk, owded, mustard yellow to orange
exture: Dry and hairless, but hairy at maturity
esh: Pale yellow and firm
talk: Tall and thin, same color as cap to own near base, yellowish veil
abit/habitat: Single to groups on ground under mixed woods
istribution: Widely noted and common
easons: Spring, summer, and fall

his bright, colorful fungus has good possibilies (in quantity) as a dye base for soft shades orange. The *Gymnopilus* **group includes 73 ecies, but none are recommended edibles.**

Pholiota squarrosoides

Cap color: Cinnamon to buff
Edibility: Edible
Size: To 11 cm. broad
Spore print: Brown
Shape: Conic to convex with maturity
Gills: White to rusty brown, notched
Texture: Viscid with dry brownish scales
Flesh: White and firm
Stalk: Tall and slim, whitish with buff-colored scales below, veil hairy or absent
Habit/habitat: Usually in large clusters on hardwoods
Distribution: Wide and common
Seasons: Late summer and fall

This fungus is commonly associated with maples, beech and birch trees, and was popular in colonial times. There are no reported toxins in this group and these large fleshy edibles usually flush in good quantity. Sauté them in butter or sunflower oil and serve with herbs and other vegetables

Mushrooms are plants that have no chlorophyll with which to make food. Like animals, ey must obtain their food from living or dead plants and animals. Imagine our natural orld without these decomposers: mushrooms, bacteria, and other fungi working to return ead substances to usable energy in order to sustain new life.

Leucopaxillus candidus

Cap color: Creamy to tawny edges
Edibility: Edible
Size: To over 20 cm. broad
Spore print: White
Shape: Broadly convex, fluted to ruffled when older
Gills: Extended down (short) stalk, crowded, sometimes forked, white to buff
Texture: Moist to dry, smooth, without hairs
Flesh: Firm and white
Stalk: Same color as cap, smooth and thick, short to 7 cm. tall, no veil
Habit/habitat: Single to several clustered or in a fairy ring in grass areas or open mixed woods
Distribution: Widely distributed in United States, infrequently seen
Seasons: Late summer and fall
Odor: Weak, like fish meal

This is a robust, fleshy mushroom with attached gills. The smooth dry caps are convex, to almost flat, to deeply ruffled and back-curled in old age. All species are found on the ground, and are possibly *mycorrhizal* **with hardwoods and conifers. There are no reported toxins in this genus, and some species are edible, though their slightly bitter taste (fresh) is disappointing.**

Leucopaxillus amarus   Ruddy pax

Cap color: Reddish brown
Edibility: Edibility unknown tastes bitter and dry
Size: To 12 cm. broad
Spore print: White
Shape: Convex to ruffled
Gills: Adnate, chalky white, close
Texture: Smooth, without hairs, and dry to powdery
Flesh: Firm and white
Stalk: Long and thick to bulbous, no veil
Habit/habitat: Single to numerous under conifers
Distribution: Widely noted
Seasons: Summer and fall
Odor: Unpleasant

This is robust fleshy mushroom with great personality but poor taste.

Leucopaxillus albissimus

Cap color: Chalky white
Edibility: Edible, tastes bitter
Size: To 9 cm. broad
Spore print: White
Shape: Convex to flat in age
Gills: Pure white
Texture: Smooth, without hairs
Flesh: Soft and white
Stalk: Long and thick with basal bulb, smooth, no veil
Habit/habitat: Single to several, occasionally in fairy rings
Distribution: Widely distributed
Seasons: Summer and fall
Odor: Sweet and aromatic

There are eight varieties of the species described by Singer and Smith (1943). In eastern North America, these mushrooms are frequently found on forest litter beneath conifers and hardwoods during summer and fall.

Entoloma abortizum Dough balls

Cap color: Pinkish brown
Edibility: Delicious, edible, tastes like yeast dough when raw
Size: To 10 cm. broad
Spore print: Salmon pink
Shape: Gilled fungi are broadly convex; the infertile "dough balls" have random shapes
Gills: Pink, close, and extend down stalk
Texture: Fertile mushroom is smooth and dry; "dough balls" look exactly like blobs of rising bread dough
Flesh: White, soft and fragile in both forms
Stalk: Whitish gray, slender, no veil
Habit/habitat: Several to numerous in cooler weather on soil and around tree stumps
Distribution: Eastern North America, gilled mushroom much less frequently encountered than its infertile counterpart
Seasons: Autumn
Odor: Of yeast bread

I was introduced to this peculiar *Entoloma* by experienced mycologist Mary Anna Wray. The infertile "dough balls" are the tasty edibles. Fried in butter or sunflower seed oil until barely translucent (three to five minutes), they make delicious "nibbles" or side dishes to accent any meal. They are exceptional at breakfast with fruit or eggs!

eter Katsaros photo

27

It is typical of the *Coprinus* at maturity to begin to dissolve as their gills produce an inky fluid. This fascinating advance of aging is caused by an enzyme within the mushroom which causes auto-digestion or deliquenscence. Among the edible species, this enzyme process is terminated by cooking.

Coprinus plicatilis

Cap color: Brown to creamy gray
Edibility: Edible
Size: To 20 cm. broad
Spore print: Black
Shape: Conic with a knob
Gills: Gray to black, turning inky (auto-digestion)
Texture: Smooth and striate
Flesh: Creamy, very thin and fragile
Stalk: Tall and slender, cap-colored, scaly and brittle
Habit/habitat: Numerous in grass during cool wet weather
Distribution: Widely known
Seasons: Spring, summer and fall

These delicate fungi are interesting to find and study, as they advance so quickly through their life cycle.

Coprinus atramentarius

Cap color: Whitish gray
Edibility: Edible
Size: To 8 cm. broad
Spore print: Black
Shape: Conic to bell-shaped
Gills: White to gray to black at maturity crowded and broad
Texture: Dry and hairless, slightly dimpled wrinkled surface
Flesh: Pallid and thin
Stalk: Tall and slender, white, dry and minute hairy
Habit/habitat: Tight clusters in grass or wood material
Distribution: Widely known
Seasons: Summer and fall

As an edible, this should not be mixed with alcoholic beverages, or nausea and gastro-intestinal upset will follow. Not everyone experiences this difficulty, but most people now know to avoid this mixture.

Gomphidius glutinosus

Cap color: Reddish brown with dark stains
Edibility: Edible
Size: To 10 cm. broad
Spore print: Smoky gray to charcoal
Shape: Broadly convex
Gills: Pale cinnamon to smoky gray, extended down stalk and close
Texture: Glutinous and smooth
Flesh: Pale buff, thick and soft
Stalk: Sturdy and short, slightly scaly below a thin ring
Habit/habitat: Several to abundant near or under conifers
Distribution: Throughout North America
Seasons: Spring, summer, and fall

These shiny, attractive grass species have several related varieties with cap color differences. They are all trusted edibles, though they are not held in high culinary esteem, tending to be a bit slimy and dark red when cooked. Nonetheless, they can often be harvested in quantities sufficient for meals and can be used to prepare a delightful casserole. This one could be confused with the poisonous *Hebeloma mesophaeum* which is very similar in appearance, habitat, and season. Both families form beneficial *mycorrhisa* with certain trees and shrubs.

Agaricus campestris
Meadow mushroom

Cap color: White to shades of brown
Edibility: Choice edible, tastes mild
Size: To 10 cm. broad
Spore print: Dark chocolate brown
Shape: Broadly convex
Gills: Free, narrow, crowded, pink to dark brown
Texture: Dry, smooth, with flattened scales sometimes
Flesh: White, thick and firm
Stalk: Sturdy, white, dry and smooth with distinctive torn ring
Habit/habitat: Several to numerous in grassy areas, lawns, and meadows
Distribution: Throughout North America
Seasons: Spring and fall, favoring cool moist periods
Odor: Mild and pleasant

This is one of our most common choice edibles. It is the one most closely related to our commercial mushrooms, *Agaricus bisporus* and *A. hortensis,* which have been propagated continually for several centuries. Called "pink bottom" when young, this agaric can be substituted freely for any conventional mushroom. I like it fried, as well as pickled, in soups, tempura and salads. This is a great trustworthy species to feed the family, and you may be fortunate enough to have it in your lawn. The rich chocolate brown spore print reassures you it is not a poisonous *Amanita,* whose spore prints are always white.

Paxillus panuoides

Cap color: Pale yellow-brown
Edibility: Non-edible
Size: To 8 cm. broad
Spore print: Buff colored
Shape: Stalkless petals or shell-like fans
Gills: Pale yellow to buff, close, radiating from point of attachment
Texture: Soft and minutely downy
Flesh: White, soft and thin
Habit/habitat: Numerous, usually overlapping on conifer logs
Distribution: Widely observed
Seasons: Spring, summer and fall

This is a small, unique family of gilled mushrooms which are not considered edibles, but contain no known toxins.

Cortinarius alboviolaceus

Cap color: Lilac to purple
Edibility: Edible
Size: To 6 cm. broad
Spore print: Rusty brown
Shape: Bell-shaped to convex to nearly flat
Gills: Pale violet to cinnamon with maturity
Texture: Dry, flattened silky hairs over cap
Flesh: Pale violet and thin
Stalk: Tall and sturdy, dry, enlarging to club shaped base, nearly cap color
Habit/habitat: Single to a few under mixed conifers and hardwoods
Distribution: Eastern North America
Seasons: Late summer and fall

These colorful fungi are a pleasure to find but I have not eaten them. *Cortinarius* **is the largest genus of gilled mushrooms in North America with nearly 600 species. Most are found on the ground and possibly are** *mycorrhizal* **with shrubs and trees.**

M. Morris photo

M. Morris photo

ortinarius cinnabarinus

color: Bright brownish red
bility: Non-edible
e: To 40 mm. broad
re print: Rusty brown
ape: Convex
s: Reddish cinnamon, numerous and close
ture: Dry and felty
sh: Brown and firm
lk: Medium and slender, cap-colored and
ny
it/habitat: Several to numerous on soil
er mixed woods
ribution: Widely known
sons: Summer and fall

s diminutive fungi favors beech/oak woods
lew England, and is often found in moss or
ong low herbs.

THE CHANTARELLES

Cantharellus cinnabarinus Red
chanterelle

Cap color: Brilliant orange-red to pink
Edibility: Edible
Size: To 40 mm. broad
Spore print: Pink
Shape: Convex-depressed to funnel-shaped
Gills: Pink to cap color, extended down stalk, narrow, forked and well separated
Texture: Dry and hairless
Flesh: White and quite thin
Stalk: Long and slender, color of cap
Habit/habitat: Solitary to numerous under hardwoods, frequently on moss
Distribution: Numerous in the east
Seasons: Summer and fall

These delicate tiny jewels are delicious when steamed with herbs and rice, or added to creamed soups. They are conversational and appetizing additions to dinner biscuits or pastries.

B.L. Wulff photo

Cantharellus infundibuliformis

Cap color: Rich brown
Edibility: Unsure
Size: To 40 mm. broad
Spore print: White
Shape: Convex and depressed with incurved margin
Gills: Lilac to yellowish gray
Texture: Hairless and dry with rough scales
Flesh: Dull yellow and very thin
Stalk: Yellow, smooth and furrowed, hollow with age
Habit/habitat: Scattered to cespitose in moss, soil or decayed wood
Distribution: Numerous and widely spread in the east
Seasons: Summer and fall

I have not eaten this species. It is closely related to several edible near-look-alikes that range widely throughout the east.

Craterellus fallax   Lavender trun

Cap color: Charcoal gray to lavender
Edibility: Choice edible
Size: To 10 cm. broad
Spore print: White
Shape: Fluted, trumpet-shaped and dee depressed
Gills: Absent, slightly striate, velvety sheen
Texture: Velvety smooth with minute hairs
Flesh: Same color as cap, thin and brittle
Stalk: Soft lavender, tubular-shaped trump
Habit/habitat: Single to several under mi woods
Distribution: Numerous in New England
Seasons: Summer and fall

This is a beautiful woodland fungus! Ligh sautéed in herb butter, this one should ke its lovely lavender coloration and be dele able served with sour cream, or cottage chee and biscuits

Craterellus cornucopioides Horn of plenty

Cap color: Dark gray-brown
Edibility: Choice edible
Size: To 6 cm. broad
Spore print: White
Shape: Trumpet-shaped, deeply depressed
Gills: Absent, slightly wrinkled ash gray to violet to black with a sheen
Texture: Slight stiff hairs and minute scales
Flesh: Dingy brown, brittle and thin
Stalk: Dull brown to shades of violet with a slight sheen
Habit/habitat: Cespitose under hardwoods in eastern North America
Distribution: Collected on both coasts
Seasons: Summer and fall

These chanterelles are among the most choice gourmet collectibles. They should be slowly simmered in butter or cream over low heat.

THE CORAL FUNGI

Clavulina cinerea Smokey coral

Cap color: Smokey gray, variable fruiting body
Edibility: Edible
Size: Ranges to 6 cm. thick and 10 cm. high
Spore print: White
Shape: Many short irregular branches
Texture: Soft, smooth and rubbery
Flesh: Grayish white and firm
Habit/habitat: Scattered on ground in woodland litter
Distribution: Widely noted and common
Seasons: Found in summer and fall

These colorful and unique fungi usually flush in good quantities. Most species are delightful edibles and can be substituted for pasta in some recipes. Young corals are the most desirable; older specimens taste a bit like stewed rubber bands. There is only one poisonous species.

Clavulina cristata — White coral

Cap color: White to creamy fruiting body
Edibility: Edible, but not delicious
Size: Grows to 8.5 cm. high and 4 cm. thick
Spore print: White
Shape: Thin, multi-branched clusters of fingers with fine tips
Texture: Soft, smooth, and rubbery
Flesh: Whitish and somewhat tough, occasionally hollow
Stalk: Club-shaped with lighter-colored base
Habit/habitat: Found singly or numerously under mixed woods
Distribution: Wide distribution
Seasons: Summer, fall and winter

These ghostly corals are exciting to find and dry, but not worth the effort as edibles.

Clavicorona pyxidata

Cap color: Tawny to rose, variable to lemon fruiting body
Edibility: Edible, tastes mild
Size: Grows to 13 cm. high and 10 cm. thick
Spore print: White
Shape: Slender, multi-branched to fine crown-like tips
Texture: Soft and pliable to tough with maturity
Flesh: White and tough
Stalk: Thick with lighter-colored base
Habit/habitat: Is gregarious in clusters or decaying wood
Distribution: Occurs widely
Seasons: Seen in spring and summer

When enough of these can be collected fresh they provide an excellent steamed dish with spaghetti sauce or butter and herb sauce – like a woodland Fettucini. These corals dry well for winter preservation. They are excellent when rejuvenated and added to oriental soups.

M. Morris photo

lavaria fusiforme   Lemon coral fungi

p color: Lemon yellow fruiting body
ibility: Edible
ze: Grows to 15 cm. high and 10 cm. thick
ore print: White
ape: Slender, cylindrical fingers, pointed at
 and originating from a common base
xture: Soft, smooth and "plastic"
sh: Yellow, tough and sometimes hollow
alk: Narrow with lighter-colored base
bit/habitat: Occur singly and scattered
der mixed woods
stribution: Common in the east
asons: Found in summer and fall

**These lovely, vivid fungi are exciting to find
d fun to deal with in the kitchen. The lemon
rals are tasty contributions to stir-fried
getables in wok cooking. They light up any
eal if delicately prepared. Do not overcook.**

THE BOLETES

**Avoid all boletes with red or orange pore
mouths (underneath sponge-like surface).
Some of the many boletes that stain blue
when touched can cause upset stomachs.
Boletes can be sautéed, broiled, grilled and
pickled. They also dry well for winter use.**

Strobilomyces floccopus   Old man of the woods

Cap color: Shaggy brown to charcoal, tints of
white
Edibility: Edible
Size: To 15 cm. broad
Spore print: Black
Shape: Broad, flat, convex
Tubes: White with large mouths, gray-brown
with maturity
Texture: Dry, rough to scaly cap surfaces
Flesh: White, staining red to black when bruised
Stalk: Long and slender, same color as cap,
shaggy, gray-white veil
Habit/habitat: Single to several on forest litter
under mixed woods
Distribution: Well known in New England
Seasons: Summer into fall

This distinctive *Bolete* **is also called the "pine
cone fungus." It dries well in nature and does
not apparently rot readily. There is not a great
taste here, nor does this species look like a
desirable edible, though it is deemed such.**

Peter Katsaros photo

Gyrodon merulioides

Cap color: Bright to dull yellow-brown
Edibility: Edible
Size: To 25 cm. broad
Spore print: Red to yellow-ochre
Shape: Broadly flat to wavy to slightly depressed
Gills: Bright to dull yellow, very shallow, bruises blue
Texture: Dry, felt-like
Flesh: Yellow, slowly changing to bluish green when bruised
Stalk: Short and slender, lateral, yellow, no veil
Habit/habitat: Single and scattered, associated with ash trees
Distribution: New England to Canada
Seasons: Summer and fall

The association with ash trees distinguishes this from other yellow boletes.

Boletus badius

Cap color: Red-brown to tawny
Edibility: Edible
Size: To 11 cm. broad
Spore print: Olive-brown
Shape: Convex to puffy-flat, like a baked cake
Tubes: Yellow changing to dull blue when bruised
Texture: Viscid in rainy weather to felt-like when dry
Flesh: White to pale pink or yellow, changing to blue when bruised
Stalk: Same color as cap or paler, reticulate thick, and sturdy
Habit/habitat: Single to several under conifers
Distribution: Principally found throughout New England
Seasons: Summer and fall

Good mature specimens resemble baked cake on the cap surface, and are thick and puffy-looking over all. This distinctive, fleshy species provides a good quantity of food. The flesh can be sliced and sautéed in margarine or butter. Pores may be removed before cooking as they tend to be slimy when cooked. Boletes make excellent additions to favorite recipes. Do not overcook.

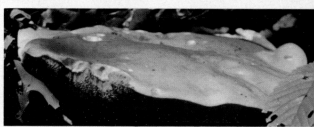

Boletus miniatoolivaceus

Cap color: Brilliant yellow with honey-colored zones
Edibility: Poisonous
Size: To 28 cm. broad
Spore print: Olive
Shape: Broadly convex to flat, surface cracked
Tubes: Pale yellow to olive, changing to blue when bruised
Texture: Dry, woolly, and cracked
Flesh: White, changing to bluish green when bruised
Stalk: Short and very fat, yellow and smooth
Habit/habitat: Usually alone or scattered under hardwoods in grassy areas
Distribution: Throughout New England
Seasons: Late summer and early fall

This showy, colorful *Boletus* is considered poisonous, and is closely related to two delicious edible varieties that are distinguished by their caps: *B. speciosus* and *B. bicolor*.

Boletus edulis

Cap color: Pale brown
Edibility: Choice edible
Size: To 25 cm. broad
Spore print: Olive-brown
Shape: Broadly convex
Tubes: Depressed at stalk and cottony white
Texture: Viscid in wet weather to dry and smooth
Flesh: White and firm, does not bruise
Stalk: Short, sturdy, and white
Habit/habitat: Usually single to scattered under mixed woods
Distribution: Throughout North America, Canada, and into Alaska
Seasons: Spring, summer, and fall

This choice edible is known as *cepe* in France, and called *steinpiltz* in Germany. Though its color and form are variable depending on habitat, it does not bruise or discolor with handling, as is so often a prominent diagnostic characteristic in this family. This *Bolete* is fabulous in soups or done with eggs and butter.

THE POLYPORES

Fistulina hepatica   Beefsteak fungus

Fruiting body: Blood red to liver color
Edibility: Good edible, tastes like raw steak
Size: To 30 cm. broad
Spore print: Pale rust
Shape: Broad, fan-shaped shelf fungi, grows in layers
Tubes: White to creamy, staining red, individual tubes
Texture: Faintly rough, moist to oozing reddish juice
Flesh: Pinkish red, thick and sturdy
Stalk: Brief thick lateral; same color as cap
Habit/habitat: Near base of hardwood stumps, usually on oak
Distribution: Widely distributed but rare
Seasons: Fall
Odor: Mild

Its distinguishing characteristics are reminiscent of raw beef, and with low heat, this *Polypore* can be treated just like beef. Use in recipes in place of beef. Sauté briefly in one tablespoon of butter and season with Worcestershire sauce or tamari sauce to taste. You will have an unmatchable main course.

Ganoderma applanatum   Artist' con

Fruiting body: Dusty gray-brown
Edibility: Non-edible
Size: To 30 cm. broad
Spore print: Brown
Shape: Thin, broad, shelf-like fan
Pores: White
Texture: Hard, smooth, sometimes cracked o looking shellacked
Flesh: Brown, cork-like, and soft
Stalk: No stalk
Habit/habitat: Single to numerous on tre trunks, stumps, and logs
Distribution: Widely noted
Seasons: Perennial

The pore surface is a noted medium fo etching pictures, which are collected an sold in giftshops. The fungus is also associ ated with heartwood rot which is infectiou and damaging to woodlands

Ganoderma tsugae — Hemlock conk

Fruiting body: Yellowish red, shellacked appearance
Edibility: Non-edible
Size: To 25 cm. broad
Spore print: White
Shape: Broad, thin, shelf-like, ridged fan
Pores: White to brown
Texture: Smooth and shiny above, smooth, felt-like and moist beneath
Flesh: White and tough
Stalk: Lateral and thick (when present), same color as cap
Habit/habitat: Single to a few on hemlock stumps
Distribution: Widespread in the east
Seasons: Summer and fall

This shiny, lovely *Polypore* is a distinctive collectible. It is sometimes found on beech trees and on other hardwoods in New England.

Daedalea quercina — Oak Daedalea

Fruiting body: White to ash gray
Edibility: Unknown
Size: Fan to 15 cm. broad
Spore print: White
Shape: Thick and fan-shaped, convex
Texture: Smooth, covered with fine short hairs
Pores: Narrow, long and irregular; usually same color as cap
Flesh: Dull white, firm but pliable, tough
Habit/habitat: Single to several on stumps and logs of hardwoods especially oak
Distribution: Eastern North America
Seasons: Perennial, lasting two to three years
Odor: Earthy, punky

This whitish, leathery *Polypore* is a perennial in our hardwood forests. It is especially found on dead oak logs, as its native name implies, and is a tenacious, ghostly decomposer.

B.L. Wulff photo

These colorful perennial *Polypores* are some of the most common and widely distributed fungi in North America. They are leathery, and are usually found in densely overlapping clusters on wood. They are stalkless, velvety and cast a white spore print. Considered survival food, extensive cooking is necessary to make them edible.

The five following polypores are common cousins in our eastern woods. They share these characteristics.

Fruiting bodies: Colorfully zoned
Edibility: Edible as survival food
Size: Fans to 5 cm. broad, except *P. squamosus* which grows much larger
Spore print: White
Shape: Thin, ruffled, overlapping fans
Pores: White and very tiny
Texture: Velvety, with a minutely hairy top surface

Flesh: Extremely thin and white
Habit/habitat: On dead hardwoods and con fers
Distribution: Most common and widely note fungi in North America
Seasons: Late summer and fall
Odor: Woodsy when wet
Taste: Tough and leathery, but palatable afte considerable boiling

---

Polyporus abietinus Blue-zoned turkey tail

A variable *Polypore* with ash-colored hairy zones, usually exhibiting a soft blue zone.

Polyporus hirsutus Amber Polypore

This smaller *Polypore* has no zones on a hairy amber, yellow, or brown cap.

Polyporus paragamenus White Polypore

This lovely white *Polypore* is found on dead hardwoods. The caps are up to 5 cm. broad, fan-shaped and ruffled, with distinct violet edges, especially underneath.

Polyporus squamosus Dryad's saddle

This large, fleshy *Polypore* (with caps to 30 cm. broad) is an edible fungi when young. It has the fresh fragrance of cucumber or melon.

Polyporus versicolor Turkey tails

The most common of the *Polypores* are the colorfully zoned "turkey tails" or "turkey wings." The thin, velvety, hairy surface shows varying degrees of white, tawny, gray, amber, buff, and black to red bands, radiating around the point of attachment.

Polyporus sulphureus Sulphur shelf

Cap color: Sulphur yellow and brilliant orange
Edibility: Delicious edible, mild taste
Size: Cluster to 60 cm. wide; cap to 25 cm. broad
Spore print: White
Shape: Fluted (ruffled), thick shelf-like fans
Pores: Bright yellow to creamy pale
Texture: Velvety smooth
Flesh: Creamy white, firm, margin soft and pliant, tough near center
Habit/habitat: Single to dense overlapping clusters on mixed woods
Distribution: Widely noted
Seasons: Summer and fall
Odor: Mild and pleasant

This one of our most exciting and intensely colored fungi. September usually brings their flush on wood, where they are known to recur every year. The edible portion is the trimmed outer margin of the young *Polypores*. As this species ages it becomes creamy white. If the young *Polypores* are sliced and sautéed in butter or sunflower seed oil until barely translucent (only two to four minutes), they should retain their excellent colors and taste. Season with tamari for a spicy touch.

HYDNACEAE The Teeth Fungi

Steccherinum septentrionale
Heartrot Hydnum

Fruiting body: Dingy buff turning yellow to brown
Edibility: Non-poisonous
Size: To 40 cm. broad
Spore print: White
Shape: Large cluster of overlapping, thick, ruffled fans
Teeth: Dull white, long and narrow
Texture: Dry with dense fine hairs
Flesh: Thick, tough and zoned
Habit/habitat: On wounds of living hardwoods, especially maples
Distribution: Eastern North America
Seasons: Summer and fall
Odor: Of ham (when dried)

The massive fruiting body ranges from 40 cm. broad to 25 cm. long, and is formed of dense, overlapping, thick clusters resembling *Polypores*. This teeth fungus is non-poisonous but is too tough and bitter to be edible. It is a most notable indication of heart rot in hardwoods.

LYCOPERDALES Puffballs

Most of the *white fleshed puffballs* are good to eat. The fruiting bodies range from the size of a small marble to that of a soccer ball, and occasionally larger. The genus forms *mycorrhizae* with higher plants, or exists as a saprophyte. It is essential to tear or slice your selected puffballs in half to be sure that the interior is consistently pure white, and that there is no form of embryonic gilled mushroom which might mean a mushroom "button" rather than a "puffball." If it is the former *do not eat it*, as it could be a poisonous *Amanita* "button." Only the young, pure white-interiored puffballs are desirable edibles. They are delicious sliced or cubed and sautéed with butter and herbs, and enhance many dishes.

At maturity the inner mass (gleba) of all true puffballs ripens to a *powdery spore mass*, which ranges in color from mustard yellow through brown to lilac. When the outer skin is pressed in dry weather, a cloud of spores "puff" from a hole in the top.

John Pawloski photo

42

ycoperdon perlatum

ap color: Dull to creamy white
libility: Edible, tastes very mild
ze: To 7 cm. thick and 7 cm. high
ores: Olive brown
ape: Pear-shaped
xture: Outer skin granular, covered with ciduous cone-shaped spines which break off aving spots
esh (gleba-spore mass): First white, maturg through yellow to olive brown
alk: Sterile base, tapering continuation of ral mass
abit/habitat: Single to many to cespitose usters in litter beneath hardwoods and conifers; metimes in moss or grass
stribution: Widely noted
asons: Summer and fall
dor: Mildly pleasing

is group is one of the safest of all edible ushrooms.

Scleroderma aurantium

Body: Yellow-brown
Edibility: Non-poisonous, tastes bitter
Size: To 12 cm. broad
Spores: Brown
Shape: Oval
Texture: Evenly warted all over
Flesh: Creamy yellow to violet-gray, nearly black and powdery with maturity
Habit/habitat: Single to several, even clustered on ground
Distribution: Widely found throughout eastern North America
Seasons: Late summer and fall
Odor: Not unpleasant

This genus is the one to avoid. The flesh is bitter, and sickness can result from ingesting it, even when cooked.

John Pawloski photo

PHALLALES
The Stinkhorns

Clathrus columnatus
Columnar stinkhorn

Fruiting Body: Coral to pale orange
Edibility: Not recommended
Size: To 5 cm. high
Spore print: Yellowish brown
Shape: Oval, chambered, stalkless
Texture: Shiny, viscid
Habit/habitat: On ground and in litter over decaying matter
Distribution: Southern New England south
Seasons: Late summer
Odor: Extremely disagreeable

This is a lacy-chambered, stalkless stinkhorn. Its coral red, finger-like projections are first clasped together at the top, and later open out like octopus tentacles. This sticky, viscid, stinky fruiting body flushes profusely on rich loamy soil beneath conifers and hardwoods.

Phallus ravenelii
Stinkhorn

Fruiting body: Greenish head on whitish stalk
Edibility: Not recommended
Size: To 16 cm. broad
Spore print: Colorless
Shape: Thick phallic stalk, with a distinct opening at the top
Texture: Slimy, sticky; stalk sponge-like and hollow
Flesh: Whitish network like a sponge emerging from an oval "egg" (volva)
Habit/habitat: Single to clusters in wood debris
Distribution: Widely found in the northeast
Seasons: Summer and fall
Odor: Nauseating, skunk-like, and carries for some distance

Flies are attracted to these fungi, as well as ants, and they walk over the sticky spore-bearing surfaces, which disperses and distributes the spores. These distinctive, odiferous fungi are common and unforgettable in their unique growth. They emerge from whitish to pinkish "eggs"; tissue-thin membrane sacs. Attached to the ground with root-like strands these fungi are fascinating to study, but not to handle!

John Pawloski photo

THE TREMELLALES
The Jelly Fungi

Auricularia auricula   Wood jelly

Fruiting body: Brown, thick, gelatinous
Edibility: Delicious, mild pleasant taste
Size: To 15 cm. broad
Shape: Ear-like lobes
Texture: Smooth and liver-like
Flesh: Thin, same color as surface, pliable, rock-like when dried
Habit/habitat: Several to many on hardwoods and conifers
Distribution: Widely collected
Seasons: Cool wet periods year-round
Odor: Pleasantly woodsy

These fruiting bodies resemble thick, dark, gelatinous ear-like fans. Favoring cool wet periods and cool nights, this jelly fungus appears in single to profuse clusters on dead wood. This is the only jelly fungus commonly eaten in New England. In the Orient, it is known as "Wood Ears" (Muk Nge) or "Cloud Ears" (Yung Nge), and is dried and packaged to be used in soups. There have been attempts to grow this one commercially in America, as is done in the Far East where logs that will produce these mushrooms at home can be purchased.

Tremella mesenterica   Witches butter

Fruiting body: Golden yellow to orange
Edibility: Not recommended
Size: To 10 cm. broad
Shape: Lobed and brain-like
Texture: Tough and gelatinous, hard when dried
Flesh: Thin and same color all the way through
Habit/habitat: Single to several on hardwoods
Distribution: Widely collected
Seasons: Summer and fall

Golden to orange lobe-like jellies on dead hardwoods. They appear singly to scattered during periods of cool weather, and are widely distributed in the northeast. This species is relatively tasteless.

Pseudohydnum gelatinosum
Teeth jelly fungus

Fruiting body: Whitish and translucent
Edibility: Edible
Size: To 3 cm. broad
Shape: Convex top on a short stalk
Texture: Jelly-like, smooth top surface with teeth beneath
Flesh: Thin and translucent
Habit/habitat: Several to numerous on decayed wood and mosses
Distribution: Widely noted
Seasons: Summer and fall
Odor: Pleasantly mild

White translucent jellies with short teeth-like spines hanging from the lower surface appear on short stalks amid moss on rotting wood. These little beauties are edible, but they barely yield a mouthful.

THE MORCHELLA
The True Morels

Morchella esculenta
The yellow morel

Fruiting body: Pale yellow to tan
Edibility: Choice edible, mild taste
Size: To 20 cm. tall
Shape: Distinct ridges and pits in hollow cap
Texture: Smooth and ridged
Flesh: Thin, pliant, same color throughout
Habit/habitat: Mixed woods, but preferring orchards and burned areas
Distribution: Widely collected
Season: Spring (brief season: two to three weeks long)
Odor: Pleasantly mild

Mycologists disagree about how many species and forms are known in this group, but they do agree that it produces the most popular and sought-after of all edible fungi in North America. These species all fruit in spring and all are edible. They are found on soil or grass under hardwoods. They favor burned areas and apple orchards, especially.

The coral-like (or brain-like) head is pitted and chambered. It is attached directly to the stalk, not skirt-like as in the false morels. Cut each morel lengthwise to examine the hollow head and stalk. Check too for insects. These delicate spring fungi are exceptional sautéed in butter and herbs, or stuffed and quickly broiled. There are numerous ways to prepare and serve the morels, and they enhance almost any food preparation!

John Pawloski photo